FAIRCHILD TROPICAL GARDEN

FAIRCHILD TROPICAL GARDEN

by

Dr. Lilly Pinkas

Photography by Dr. Joseph Pinkas

Library of Congress Catalog Card Number : 96-69024

ISBN 0-9652810-0-0

Edited by Lenka Pinkas Wagner

Designed by Panorama Foto-Druck-Verlag Lenka Wagner

Printed in the United States of America
by Hallmark Press, Miami, Florida
First Hardback Edition May 1996

ACKNOWLEDGEMENTS

I would like to express my grateful appreciation and sincere thanks to Don Evans for his kind help on many aspects of this book and for his infinite patience in answering the multitude of questions I have had while working on this project. His expertise was truly invaluable. A very special thanks to Dr. Brinsley Burbidge for his respected advice and for giving his time freely especially while reviewing the manuscript. Many thanks to Craig Allen for his help with plant identification. I would also like to thank many other FTG staff members for their contributions which made my work easier. A special thanks to Jan Pinkas for his help with graphics.

Fairchild Tr

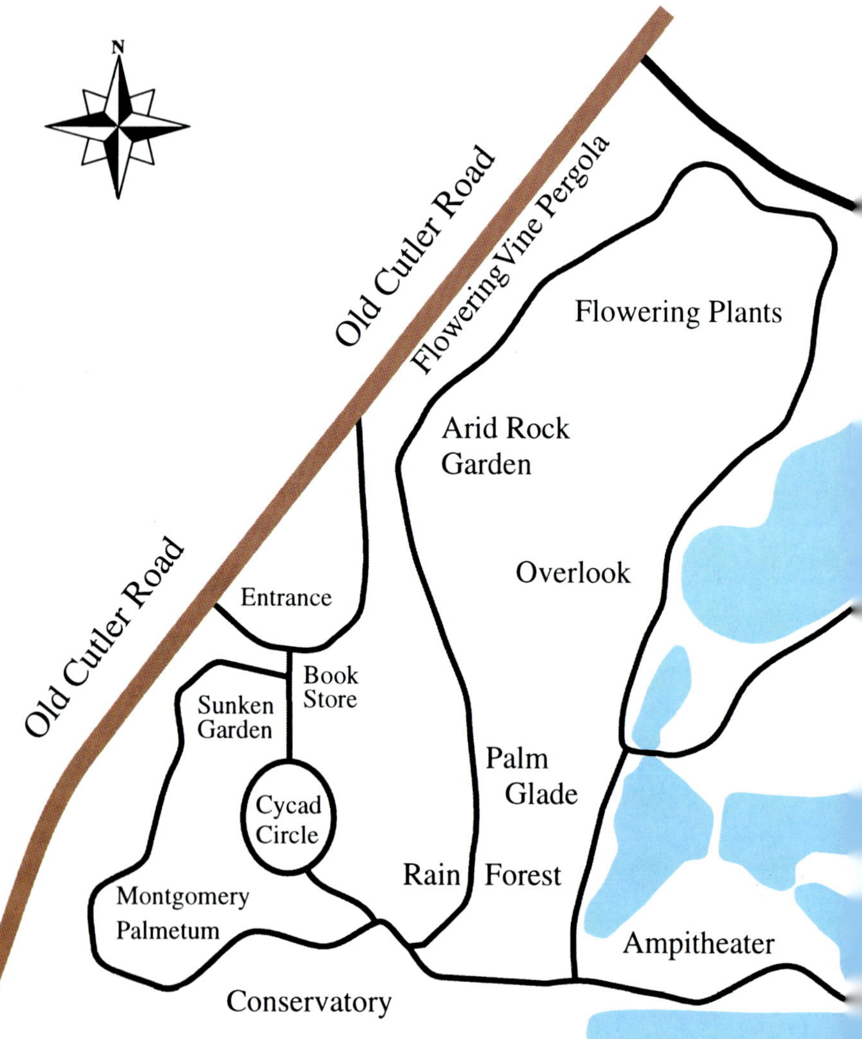

ical Garden

Matheson Hammock Picnic Area
Mangrove Preserve
Chachi House
Bahamian Plants
Hibiscus Garden
Bird Habitat
Coconuts

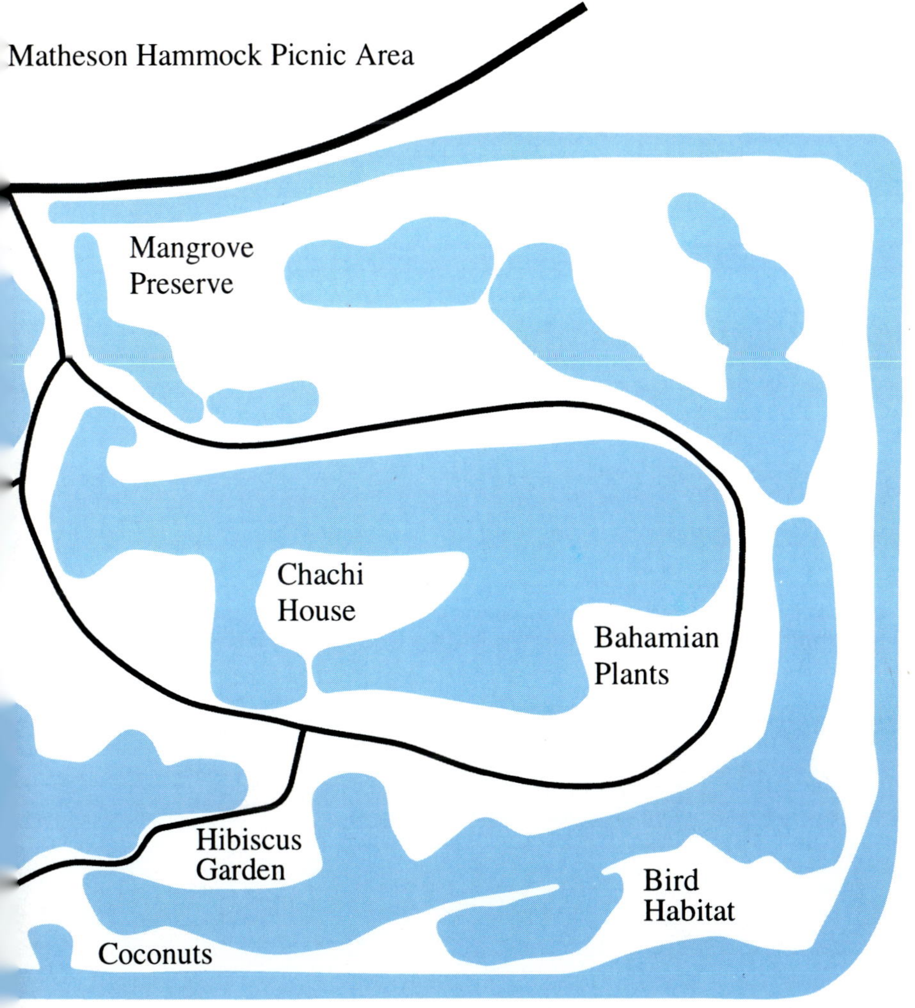

Introduction

Located just a few miles from downtown Miami, Fairchild Tropical Garden appears a tropical oasis amidst the bustle of Southern Florida. Spread out over 83 acres is the largest collection of palms, cycads, tropical and subtropical plants in the continental United States.

In the 1930s Colonel Robert H. Montgomery, an avid amateur plant collector, envisioned establishing a tropical garden in Southern Florida, the only region in the continental United States where it is possible to grow tropical and subtropical plants throughout the year. Since his own palm and cycad collection had generated much interest, he believed a tropical garden would appeal to and be supported by the public.

Colonel Montgomery was encouraged by, among others, his close friend, Dr. David Fairchild, for whom the Garden was later named. One of the most noted and respected American plant explorers of his time, Dr. Fairchild enjoyed worldwide recognition and respect in the field of botany. He not only created and headed the U.S. Department of Agriculture "Seed and Plant Introduction Section" but also introduced over 2,000 varieties of plants, fruit, trees, grains and vegetables into this country. It is quite amazing to realize that many basic staples present on an American table today were his introductions just 60 to 80 years ago: dates and cotton from Egypt; rice, soybeans and cherry trees from Japan; wheat from Russia; cauliflower from Italy and polyembryonic mangos from Indonesia, just to mention a few.

The renowned Harvard-educated landscape architect, William Lyman Phillips, was retained to create the Garden's design. At the time he was superintendent of the Civilian Conservation Corps in Southern Florida, which later did much of the original excavation and construction.

The layout of the Garden is based on ecological communities - plants which share the same habitat - as well as on taxonomic plots - which include plant species belonging to the same families. The result is a classic architectural design where wide open spaces, lakes, and plant groupings provide variety and diverse scenery throughout the Garden.

After several years of planning Colonel Montgomery was able to realize his dream when Fairchild Tropical Garden opened in 1938.

Phillips Gate, the original entrance to the Garden

Since its inception the goals of the Garden have been not only to display tropical and subtropical plants but also to promote botanical research and the conservation of rare species as well as to educate both scientists and the public.

Winding through the Garden's 83 acres are pathways leading through the Garden's various sections. Many visitors choose to start their exploration by taking a guided tram tour of the Garden to provide them with an initial overview. The trained guides combine botanical and historical information in the approximately two mile long tour. Or visitors should feel free to explore at their own pace and leave the beaten path, as long as they don't pick flowers or seeds. Smell the flowers, touch the leaves and bark of unusual trees, or just admire the surrounding beauty.

Garden's Tramstar

Heliconia (Heliconia humilis)

Reflection of palms in lake

The Garden

Fairchild Tropical Garden is home not only to the largest **Palm Collection** in the continental United States but even boasts one of the largest and finest palm collections in the world. The Garden's original palm collection was located primarily in the Palmetum, but today palms can be found throughout the entire Garden.

Of the more than 2500 known species of palms in the world, Fairchild Tropical Garden's living collection includes more than 550 of them. Palms are mainly tropical or subtropical plants. Somc palms grow as shrubs or even vines climbing into the tree canopy. The diversity in the sizes and shapes of various species of palms is truly amazing. Just consider the following:

- The trunk of palms may grow only several inches or almost 200 feet high.
- The diameter of palm trunks may vary from 1/4 inch to three feet.
- Palm fronds range from just a few inches to more than 30 feet in length.

- Palm seeds may be smaller than the head of a match or weigh as much as 40 lb.
- The trunks of palms are most often solitary but may be branched, as in the case of the **Doum Palm** (*Hyphaene thebaica*).

Queen Palm (Syagrus romanzoffiana)

The number of individual palms of a given species may be many or very few. Some palms still occur in great numbers, one example being *Copernicia alba* that grows in solid stands on hundreds of square miles in an area where Paraguay, Brazil and Bolivia meet. It is estimated that this region alone may contain 500 million palms.

Bottle Palm (Hyophorbe lagenicaulis)

Some species of palms are severely endangered or almost extinct. The **Bottle Palm** (*Hyophorbe lagenicaulis*), a feather-leaved palm with a swollen trunk, is native to the Mascarene Islands in the Indian Ocean and once covered large areas of the islands. Now bottle palms are nearly extinct in the wild.

Another example of an endangered species is the **Sargent's Cherry Palm** (*Pseudophoenix sargentii*). A native of the West Indies, this palm used to be quite common on Long and Elliot Keys in Florida. In the early 1900s they were almost wiped out by collectors of landscaping plants as well as by the destruction of native hammocks to provide land for pineapple and lime groves. Seeds collected from palms on Long and Elliot Keys were brought to the Fairchild Tropical Garden; palm seedlings were later reintroduced back to Long and Elliot Keys.

Several of the palm species are of great economic importance. For example, the **Coconut Palm** (*Cocos nucifera*) is without question the best known as well as one of the most beautiful. The sight of coconut palms is synonymous with the tropics.

Coconut Palm (Cocos nucifera)

The Coconut has been cultivated for so many centuries that its origins in the wild are not known. Coconut palms grow - oblivious to salt water - in sandy soils along shorelines in the tropics.

Buoyed by their husks, coconuts float extremely well and are frequently carried by ocean currents to distant regions. Even after floating in salt water they retain their germinating power for several months and sprout easily. To most of us, coconut palms used ornamentally provide tropical grace to our garden landscape, but around the world many people depend on the various palm products. The coconut palm is probably one of the most versatile palms. The husk of the coconut is the source of coir used for mats and ropes. The raw flesh of the coconut is used for food or, if dried, is called copra, a great source of vegetable oil. The hard inner shell of the nut is used to make bottles and cups but also used as a fuel or to make charcoal. Coconut milk is used in cooking or as a beverage. Sap obtained by tapping the flower stalk is a source of sugar, alcohol or vinegar. Trunks of coconut palms yield wood for use in construction or the making of furniture. Palm leaves are used for baskets or roofing material. Hundreds of other uses could still be mentioned since every part of the palm is used. According to an old Indian saying: "If you take care of a coconut palm for the first seven years of its life, it will take care of you for the rest of your life".

Triangle Palm (Dypsis decaryi)

The **Date Palm** (*Phoenix dactylifera*), native to the Middle East, has been cultivated for fruit since about 6000 B.C. This tree requires very little water. It starts to bear fruit at five years and may continue bearing fruit for over 100 years. One date palm may bear up to 500 lb. of dates. All parts of the tree are used, and at least 800 uses have been counted. Its edible fruit is a major crop of several Middle Eastern countries; it is estimated that Iraq alone has 20 million date palms. Dates have been grown commercially in California since 1890, and today their production - combined with that of Arizona - reaches several million pounds a year. The climate of South Florida is, however, not suitable for the production of good tasting edible dates.

Palms are also important in sugar production. Sugar is produced not only from the **Sugar Palm** (*Arenga pinnata*) but also from several other species such as the **Nipa Palm** (*Nypa fruticans*), the **Fishtail Palm** (*Caryota urens*), the **Talipot Palm** (*Corypha umbraculifera*), the **Wild Date Palm** (*Phoenix sylvestris*) and even the **Coconut Palm** (*Cocos nucifera*).

As in the case of the **Palmyra Palm** (*Borassus flabellifer*), the flow of sap is obtained by tapping the flower stalk. This will start the flow of a few quarts of sap daily lasting for several weeks.

Oil palms are invaluable as producers of oil for a multitude of uses. Oil of the **African Oil Palm** (*Elaeis guineensis*), native to West Africa, is a valuable commodity, and one of the important export products of this region of Africa. The worldwide production of oil from African oil palms amounts to millions of tons. **American Oil Palm** or **Cohune Palm** (*Attalea cohune*) and **Scheelea** (*Scheelea zonensis*), both native to Central America, provide oil from their seeds which is used for cooking, soap manufacture, etc.

And what about the **Seychelles Palm** (*Lodoicea maldivica*) that produces the largest seed in the world? The endangered palms grow on only three of the islands in the Seychelles, a group of islands in the Indian Ocean. The Seychelles palm is frequently called coco-de-mer or double coconut because the nut looks like two large coconuts joined together. The nuts weigh 30 - 45 lb., and there may be as many as 60 to 70 such nuts on a single tree. This large fan palm grows about 90 to 100 feet tall. It may flower when it is 20 to 30 years old but does not mature or bear fruit until at least 100 years old.

Canary Island Date Palm (Phoenix canariensis)

Trunks of Palms

Bismarck Palm (Bismarckia nobilis)

Another fascinating palm is the **Talipot Palm** (*Corypha umbraculifera*), native to India and Sri Lanka. This palm blooms and bears fruit only once in its life, usually when 30 to 80 years old, and then begins to die. The Talipot palm produces the largest inflorescence in the plant kingdom; its length may reach up to 35 feet, containing as many as 10 million flowers and producing up to a ton of seeds.

Consider the amount of information you have learned about just eight of the Garden's approximately 550 palm species. The remaining 542 (or so) palm species are just waiting to be discovered by you. Come visit, learn more, read. You will not be disappointed.

Solitude

Cycad (*Encephalartos ferox*)

Cycads are members of an ancient family, survivors of plants known to have existed in the Mesozoic Era from about 225 million to 65 million years ago in the age of dinosaurs. Distributed in tropical and subtropical regions they represent the most primitive, oldest living forms of seed-bearing plants. Cycads are palmlike, woody plants which grow very slowly. The stem contains a large pith surrounded by a fairly narrow zone of soft wood. The stems of some cycads are partially buried or totally underground. Cycads are strictly dioecious, meaning that male and female reproductive organs are on separate plants. Male cones produce pollen which is then transmitted by wind to female cones where fertilization occurs.

Fairchild Tropical Garden has one of the largest and most extensive collections of cycads in the world. Of the world's 200 known species of cycads, the Garden boasts 180 species. The Garden's cycad collection is located not only in Cycad Circle but throughout the upland area as well.

In our part of the world cycads are widely grown as ornamental plants. In other parts of the world, young leaves and seeds of some species are consumed. Starch obtained from the powdered pith of some cycads is also edible after the toxic alkaloid (cycasin) is removed either by thorough washing or by curing its mashed pith in the sun for long periods of time. When ground, this powdery "flour" was used to prepare bread. **Florida Arrowroot** or **Coontie** (*Zamia pumila*) is a small cycad with a short trunk, all of which is underground. Its starchy pith was a major food of Florida's Seminole Indians; the harvest took place in the winter months when the starch content is highest.

Cycad (Dioon edule)

Lily (Hymenocallis latifolia)

Keep in mind that the **Sago Palm** (*Cycas revoluta*), a favorite landscape plant in South Florida, is not really a palm but a cycad. The "real" Sago Palms from which the sago starch is commercially produced are *Metroxylon rumphii* and *Metroxylon sagu*, both native to Indonesia.

Garden scene

The hot, steamy and humid environment of the equatorial rain forest is simulated in the **Rain Forest** exhibit. The high canopy is provided by **Live Oak** trees (*Quercus virginiana*) and various tropical tree species. A multitude of smaller understory plants such as huge epiphytic **Staghorn Ferns** (*Platycerium bifurcatum*), orchids, bromeliads and aroids grow on the ground and also high in the branches of trees.

The beauty of a tropical rain forest with its multitude of plants and organisms is sure to captivate and amaze us, but at the same time we must work to prevent further rain forest destruction. The rain forest is being cut down and destroyed at an alarming speed, whether timber harvesting, agriculture or just "progress and development". The consequences are soil erosion, climate changes and a permanent loss of animal and plant species.

Can a tropical rain forest regenerate? If it can, how long would it take? Nobody knows the answers! Let's not wait until it's too late.

Above: Red Ginger (Alpinia purpurata),
Below: Bromeliad (Neoregelia hatchbackii)

Conservatory (Windows to the Tropics)

A beautiful, new, state of the art **Conservatory** was opened in the spring of 1996. This exciting structure stands on the site of the old Rare Plant House, which sustained severe damage by Hurricane Andrew in 1992.

In addition to simply displaying rare plants in their natural setting, the Conservatory plays an important role in the Garden's conservation efforts. Many of the plants here cannot be grown outdoors in Southern Florida. Some of the specimens are too tender to grow in exposed outdoor locations, some are cold sensitive, others may need more shade, special soil or increased humidity. Any of these conditions can be easily controlled and adjusted in this facility.

And the Conservatory is ideally suited to provide an educational experience as well. Featured is a Windows to the Tropics exhibit highlighting the plant life in tropical environments. The collections are arranged along various themes, such as plant/animal interactions, new plant introductions, plants and people, etc.

Cattleya Orchid (Hybrid)

Cattleya Orchid (Hybrid)

Moos Memorial Sunken Garden was created out of what was originally just a sink hole. However, instead of filling the hole - as was initially contemplated - a small oasis of beauty and solitude was created. Each entrance is guarded by a **Live Oak** tree (*Quercus virginiana*). As one descends a gently sloping path carved out of limestone, suddenly a waterfall and a small pool appear. Tall palms tower around the upper walls; anthuriums, philodendrons, ferns and begonias grow out of rock crevices and lower walls.

Vanda Orchid (Hybrid)

Flame Vine (Pyrostegia venusta)

One of the Garden's landmarks, the **Flowering Vine Pergola** stretches 560 feet along the Old Cutler Road wall. The stone pillars are dominant features of this section of the Garden. Constructed by members of the Civilian Conservation Corps just before World War II, the original pillars of the pergola were built out of oolitic limestone. A great number of flowering vines provide a beautiful display of colors throughout the year.

Traveller's Tree (Ravenala madagascariensis)

The **Traveller's Tree** (*Ravenala madagascariensis*), a truly remarkable relative of the banana family, is much cultivated throughout the tropics. In Florida the traveller's tree is grown ornamentally. In its native Madagascar there are many uses for this tree. The wood is used for house construction, leaves are used for roofing, the sap yields sugar and even its seeds are edible. Leaves up to 10 feet long grow in a flat plane and form a fan-shaped cluster on top of the palm-like trunk. It is said that the fan always points in a north to south direction like a compass. This is, however, not true and depending on how the tree is planted it may point in any direction. The tree received its common name because of its reputation as a source of water for thirsty travelers. A closed cavity at the base of each leaf may hold a quart or more of water; a hole can be drilled in the base of a leaf stalk to obtain the water.

Palms in the Lowlands

Another of the fascinating trees one can admire in the Garden is the **Baobab Tree** (*Adansonia digitata*). Native to Africa, it is one of the largest trees in the world. The tree was named after the botanical explorer Michael Adanson who discovered it in 1794. It is not uncommon for the barrel-like trunk to reach a diameter of 30 feet. Trees with a 100-foot girth are known to exist, although the height of the tree rarely exceeds 60 feet. The Baobab quite often looks even fatter than it actually is because a tree 40 feet tall may have a trunk 30 feet wide.

The age of some of the larger trees is estimated at 5000 years, making it one of the longest lived. The Baobab is so extraordinarily shaped that an Arabian legend has it that "the devil plucked up the baobab, thrust its branches into the ground and left its roots in the air". The tree has adapted to a dry climate marvelously, developing a very deep root system, large branches that bear leaves only during the wet season and a pithy trunk with very little wood but great water storage capabilities. A large tree may store several thousands of gallons of water in its fibrous trunk.

Just about all the parts of the baobab tree are put to good use. A very strong fiber from the bark is used to make rope and cloth. Furry fruit up to one foot in diameter hang on long stems. They are sought by men as well as by monkeys for its cool and tasty pulp. Leaves are used for leaven or as a vegetable. Quite frequently a fungus causes the trunk of the baobab tree to become hollow; sometimes trunks are purposely excavated. The hollow trunks are used as temporary shelters or water reservoirs in some parts of Africa. In times of drought elephants often rip open the trunk of the tree to get the water out of its pulp.

One of the Baobab trees growing in Fairchild Tropical Garden was planted in Dr. David Fairchild's honor during the dedication ceremony on March 23, 1938. The tree came from a seed Dr. Fairchild himself collected while on an African expedition in 1927.

Giant Shaving Brush (Pseudobombax ellipticum)

In March the **Giant Shaving Brush Tree** (*Pseudobombax ellipticum)* is an explosion of color against a blue sky. Flower buds resemble dark brown, cylindrical acorns a few inches in length. The buds spring open at night, and the next morning petals are curled back exposing numerous long, pink stamens. The flowers look like colorful shaving brushes fastened to the tips of leafless branches. This tree, native to Mexico, is considered one of the most beautiful flowering trees.

Giant Shaving Brush (Pseudobombax ellipticum)

Cannonball tree blossom (Couroupita guianensis)

It is quite rare to see a blossoming **Cannonball Tree** (*Couroupita guianensis*) in the continental United States, and the one at Fairchild Tropical Garden is especially beautiful. The cannonball is a timber tree native to Guyana but in other tropical countries is cultivated for its unusual flowers and fruit. The pleasantly fragrant flowers are pollinated by bats in the wild, but here they must be hand-pollinated. Flowers emerge on special floral branches growing from the lower trunk. The fruit of the cannonball tree is four to eight inches in diameter, chocolate brown and quite inedible. The pulp of the fruit is full of seeds and has an unpleasant odor when ripe. It is a sight to see these cannonballs hanging on heavy strings suspended from the trunk of the tree. On a windy day the pounding of hard-shelled cannonballs against the trunk and each other can create a quite noisy affair.

The **Sapodilla tree** (*Manilkara zapota*) at Fairchild Tropical Garden is probably one of the largest in this country. Many Garden visitors welcome the pleasant shade it provides directly in front of the Rain Forest Cafe.

Sapodilla is native to Central America and Mexico where whole forests grow on limestone soil. In tropical America the tree is cultivated for its fruit which is deliciously sweet and juicy. Sapodilla trees are tapped to obtain the milky sap - the main source of chicle originally used in the making of chewing gum.

Bromeliad (Guzmania sanguinea)

The Lowlands section of Fairchild Tropical Garden comprises a 57-acre area. These lowlands were once marine flats before the Bay and marshes were filled. The creation of lakes not only provided additional fill for this area but, maybe even more importantly, created permanent open spaces all in accordance with the Garden's grand design.

Several banyan trees grace Fairchild Tropical Garden. The **Banyan tree** (*Ficus benghalensis*) is an evergreen tree originating from India where it is sacred to the Hindus. A banyan is a fig tree that has developed numerous aerial roots which grow into the ground from its horizontal branches. These roots eventually grow to become auxiliary trunks helping to support the branches and allowing the tree to grow outward. "Extra" trunks of the banyan may number in the hundreds. In this way, some trees spread out over a significant area. There are known instances where a single tree covered several acres of land. In the Calcutta Botanical Garden there is a banyan tree that is so large and has so many trunks that it takes at least 10 minutes to walk around it. There is yet another enormous specimen in India that measures more than 2000 feet in circumference.

Decorative Pineapple (Ananas bracteatus)

The shores of Pandanus Lake are lined with **Screw Pine** (*Pandanus*). Pandanus gets its common name from the spiral screw-like setting of its leaves and the pineapple-like appearance of its fruit; however they are not related in any way to pine trees. Pandanus is native to Southeastern Asia and the islands of the Pacific. There are many known Pandanus species as well as many more not yet identified. Sometimes called walking trees, the many cylindrical aerial roots growing at oblique angles out of the leaning trunk make them appear as if they are walking on stilts. In many tropical Pacific regions Pandanus has been one of the most important resources for natives since ancient times providing food, clothing and shelter.

Mangrove Preserve

The mangrove family, quite common on tropical and subtropical coastlines, includes many species of trees and shrubs. South Florida is an ideal region to study mangroves since the three main species are all represented here and grow in many different soil types: **Red Mangrove** (*Rhizophora mangle*), **Black Mangrove** (*Avicennia germinans*) and **White Mangrove** (*Laguncularia racemosa*).

One may not realize that large parts of Fairchild Tropical Garden were originally a mangrove swamp. Fairchild Tropical Garden maintains an original stand of mangroves in the lowlands where they are preserved in their natural state.

Red mangroves are not only salt water tolerant, they seem to thrive in it. They grow along the shoreline and well out into the mudflats. Mangroves play an important role in protecting and stabilizing the shoreline. Decomposing leaves caught in the dense tangle of aerial roots create new land. The mangroves also help to support a rich marine ecosystem by providing food supply, nursery grounds and sanctuary for a great variety of marine life, especially juvenile fish species and shrimp.

Black mangroves fairly typically grow further inland, covered by waters of high tide but quite exposed during low tide.

White mangroves usually grow further away from the water, frequently behind other mangroves.

At times all three mangroves may be found growing in mixed stands.

Bahamian plants

A section of the lowlands was reserved to establish a collection of rare Bahamian trees and shrubs, some of which are found only in the Bahamas. Considering the continuing development of the Bahamas, it is quite conceivable that some of these plants might one day become endangered or even extinct. All of these plants grow quite well in our South Florida climate which is very similar to that of the Bahamas. Propagated and maintained in Fairchild Tropical Garden, this collection is readily available for scientists to study (without having to visit several Bahamian islands). A few plants of the same species are planted so that if some are lost, the species will still be represented or can even be reintroduced back to the Bahamas.

The **Migratory bird habitat** is a relatively new addition located in the Garden lowlands. South Florida and the Florida Keys lie along a major flyway for birds migrating to the Caribbean and South America. Canopy trees were planted to provide cover, shelter and food. Future plantings of several additional understory plants will complete this project. This habitat will serve as an extension of Florida

Keys natural feeding areas for migratory birds and also as an area where people can see native South Florida trees and plants.

Chachi House

Chachi House

The newest addition to the Garden lowlands is a Chachi house as a part of the "People of the Rain Forest Exhibit". The staff of the Fairchild Tropical Garden wanted to demonstrate what life in a tropical rain forest is like, and the Chachi people warmly embraced the idea.

This exhibit is meant to educate us not only about the indigenous people of the tropical rain forest in northwestern Ecuador but also about the tropical rain forest itself. There are only about 4000 Chachis left. As the rainforest is destroyed, their entire existence is in danger of extinction. The Chachi people are under immense pressure to cut and sell their forest; the same forest that allows them to be self-sufficient and provides them with just about everything they need: food, water and shelter. To grasp the tremendous importance of this highly endangered habitat, one must realize that the remaining tropical rain forests now occupy only a few percent of the earth's surface, yet they still account for more than half of the world's plant and animal species.

Three Chachi men came to the Garden and built a true replica of a Chachi house. It is a house on stilts, constructed from bamboo and palms with a thatched palm roof and without walls. Not only were all of the building materials brought from Ecuador, but the exhibit is complete with dug-out canoes, furnishings and even a garden.

Hibiscus (Hibiscus rosa-sinensis)

The Hibiscus Garden

The Fairchild Tropical Garden hibiscus collection is not to be missed. One can wander in amazement throughout the multitude of colors of hibiscus blossoms. Origins of today's hibiscus can be traced to the **Chinese Red Hibiscus** (*Hibiscus rosa-sinensis*). Today there are several thousand varieties and hybrids of hibiscus. Blooms on most last just one day; hybrids have been developed which usually bloom for two days.

Hibiscus (Hibiscus rosa-sinensis)

A **Bamboo garden** was just recently established in the lowlands of Fairchild Tropical Garden. Bamboos are grasses; some have woody stems and may range in height from a few inches to more than 100 feet. Dr. Fairchild was fascinated by the beauty and economic value of bamboo. Many kinds of bamboo are cultivated for their ornamental value, yet very few plants offer more practical uses than bamboo. The young shoots and seeds of some bamboos are edible. The wood is used for a multitude of building materials; furniture, mats and paper are just a few of the additional uses.

The **Sausage Tree** (*Kigelia pinnata*) is indeed an oddity in the plant world. The tree is native to tropical West Africa and is cultivated as a curiosity for its strange looking fruits resembling sausages dangling from its branches. The "sausages" weigh five to 12 pounds, take a year to ripen and are not edible. However, the Masai people of Africa use the "sausages" to make an alcoholic beverage for use in their festivities and celebrations. "Sausages" are preceded by night blooming, unpleasant smelling, velvety deep red flowers, usually opening one at a time. In their natural habitat they are pollinated by bats, but in the Garden they must be hand-pollinated in order to assure another crop of "sausages".

Sausage tree (Kigelia pinnata)

Water lilies

Conclusion

We hope that the preceding pages and pictures demonstrated the unique place Fairchild Tropical Garden holds among the world's leading botanical gardens. A 1992 statement adopted by the Fairchild Tropical Garden Board of Trustees best summarizes the missions and goals of the Garden:

"- To be a premier tropical botanical garden of the world;

- To set the highest possible standard in landscape design and exhibitions, living collections, and horticultural practices;

- To be a primary source of information on tropical plants through research and education;

- To inspire positive attitudes and behavior toward the urban and natural environment."

Such goals may seem outright overwhelming; it is certainly no small task to accomplish them. Fortunately the Garden's staff of horticulturists, scientists, curators and educators is actively supported by more than 7000 Garden members and hundreds of dedicated volunteers giving their time, talent, sweat and ideas in order to accomplish the common goal.

The Garden's Montgomery Library collection of about 8000 volumes is a remarkable treasure of horticultural and tropical botany books valued by researchers, scientists and students from around the world. The Garden's impressive herbarium collection includes more then 75,000 dried plant specimens for research.

Students at Garden

The institution plays an extremely important role in education. From grade school pupils and their teachers, the general public, university students or professional botanists - the entire spectrum is covered. Whether it is an informal instruction, garden tour, educational workshop or a formal seminar, it is carried out knowledgeably and with enthusiasm. In the long run education is the key! Schoolchildren and adults alike are introduced to nature, plants and conservation. Most importantly, though, is the increased awareness that what we do and how we behave can and will make a difference.

It is projected that 20 percent of the world's biodiversity will be lost in the next 25 years with most of this loss occurring in tropical regions of the world. With this in mind, there is a sense of urgency to collect and preserve the plants that may one day be the last living specimens of their kind. At the time of this writing, the Garden had over 9,300 plants, representing more then 3,060 species, with approximately 1,000 new accessions made since Hurricane Andrew.

And let's not forget the research and science at Fairchild Tropical Garden. The Garden has varied programs in the plant sciences and in plant conservation. Staff botanists are actively studying rain forests in the West Indies and Central

America, learning new ways to classify palms worldwide, describing the effects of fire on Florida's native cycad. They are trying to understand how water flows through the stems of tropical vines and to develop better ways to prune and grow tropical fruit trees. The Garden's scientists study the life cycle of rare plants of South Florida, advise local governments on how to manage the few remaining natural areas and publish reference books and technical papers about tropical plants. The Garden's scientists are also mentors, advising students at local universities and teaching tropical botany to school teachers and the public at large. In their search for new specimens, they are also plant explorers, bringing back seeds and cuttings of new plant introductions.

The Fairchild Tropical Garden, as a member of the Center for Plant Conservation, has been designated a national site for endangered species of South Florida and the West Indies. And stressing over and over again, we have to realize that we must balance and redefine our relationship with nature or many species and endangered ecosystems will be lost forever. Keeping all of the noble goals in mind, let us not lose sight of maybe the most important fact that the Garden is a place of enjoyment and great beauty.